IMAGES
of America

LEBANON

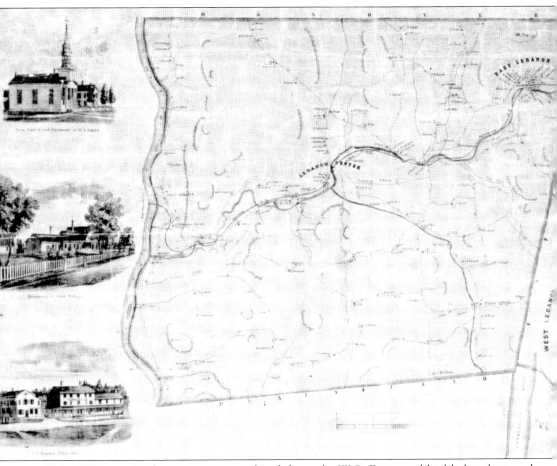

This 1855 map of Lebanon was surveyed and drawn by W.C. Eaton and highlights the town's major villages: West Lebanon, "Lebanon Center," and East Lebanon. The central village and West Lebanon were also featured in detailed views to the right of the map, opposite the illustrations of notable landmarks to the left. (Dartmouth College Library.)

ON THE COVER: Two department of public works employees pose with their municipal trucks in front of the Lebanon town sheds on Spencer Street in 1945. A fire destroyed these buildings five years later, but the facility was rebuilt on the same site. (Lebanon Historical Society.)

IMAGES
of America

LEBANON

Nicole Ford Burley

ARCADIA
PUBLISHING

Copyright © 2023 by Nicole Ford Burley
ISBN 978-1-4671-6045-2

Published by Arcadia Publishing
Charleston, South Carolina

Printed in the United States of America

Library of Congress Control Number: 2023936139

For all general information, please contact Arcadia Publishing:
Telephone 843-853-2070
Fax 843-853-0044
E-mail sales@arcadiapublishing.com

Visit us on the Internet at www.arcadiapublishing.com

*For Lebanon's residents, past and future, and
for all who know and love the city.*

CONTENTS

ACKNOWLEDGMENTS

This book owes its greatest debt to the past curators and collectors of the Lebanon Historical Society. Their commitment to preserving artifacts and photographs has built a rich visual archive of Lebanon's history, only a small portion of which is reproduced in this book.

I am also grateful to the Archives de la Ville de Montréal, Jessica Bright at the Bryn Mawr College Special Collections, the Dartmouth College Library, Kevin White from the Historical Marker Database (hmdb.org), the Keene Public Library, Library and Archives Canada, the Library of Congress, the National Archives, the National Library of Medicine, the National Portrait Gallery–Smithsonian Institution, the Vermont Historical Society, and Dominic Hall at the Warren Anatomical Museum for generously sharing images from their collections for this project.

Finally, I thank the members of Lebanon's Heritage Commission and the Lebanon Historical Society for their support, insights, and invaluable contributions to this project. I am particularly grateful to Richard Ford Burley, Stephanie Jackson, Barbara Krinitz, Jerry Rutter, Matt Smith, and Devin Wilkie for their careful review and excellent feedback, and to Phil Edson for his assistance with all things West Lebanon.

Unless otherwise noted, all images appear courtesy of the Lebanon Historical Society.

INTRODUCTION

Lebanon is the largest city in the Upper Connecticut River Valley and a major commercial and industrial hub in the region. However, Lebanon, situated as it is in a region responsible for remarkable artistic, manufacturing, and cultural contributions, is often overshadowed by its neighbors. Lebanon's educational and intellectual contributions are overshadowed by neighboring Hanover and its showpiece Dartmouth College. Histories of furniture manufacturing and agricultural innovation focus on the Shaker community in Enfield rather than on Lebanon's craftspeople and inventors. The artistic contributions and larger-than-life personalities of the Cornish Colony 10 miles to the south rise above Lebanon's own.

It is Lebanon, however, that served as the western terminus of the Northern Railroad when the first rail lines stretched northwest from Concord. Lebanon was the crossroads of two of New Hampshire's earliest turnpikes and a major stop on the interstate highway system. No other community can lay claim to Lebanon's unique collection of people, including the founder of a major American religion, a noted artist of the White Mountains, the namesake of Fort Bliss, and one of psychology's greatest subjects. Lebanon was unparalleled amongst its neighbors for its manufacturing and mill industries, and after their decline, Lebanon would lead the region in scientific and medical innovation. It is Lebanon that would host one of the country's most successful pedestrian malls, constructed amidst the ashes of the city's second great fire.

Though the history of the modern settlement founded by European colonists begins in 1761, the history of Lebanon started much earlier. Lebanon's first inhabitants likely came to the area around 11,000 BCE as retreating glaciers exposed fertile land and left behind large glacial lakes. Paleoindians lived in the Connecticut River Valley and throughout New England for millennia, developing from a predominantly nomadic hunting society into a more settled farming and agrarian culture. By 800 CE, Lebanon was home to the Western Abenaki, providing fertile farmland and bountiful wildlife that the Abenaki utilized seasonally. The land that would become Lebanon was well-suited for hunting and farming in the warmer months, providing resources for the community to use throughout the winter in villages like those of Koas (near Newbury, Vermont) and Squakheag (Northfield, Massachusetts). The Abenaki continued to live in Lebanon into the early modern era, when they were forced to flee or assimilate following the arrival of European colonists.

The town of Lebanon was created by a charter from the Royal Governor Benning Wentworth in 1761, and European colonists began to arrive in Lebanon the following year. The first years were dedicated to clearing land, constructing houses and mills, and organizing a town spread across several miles of challenging terrain. Within 20 years, however, the settlement had grown from a handful of enterprising men to a town of 1,190, with several commercial and industrial centers and a distinguished new meetinghouse.

From this promising start, Lebanon continued to grow and expand. The community constructed a network of roads connecting Lebanon's principal villages—East Lebanon, the central village, and West Lebanon—and the neighboring communities. Lebanon's residents embraced the advent of the railroad in the 1840s, capitalizing on the rapid transportation to move not only people but also manufactured goods to the largest commercial hubs of the region. Accordingly, Lebanon's manufacturing industry flourished, harnessing the power of the Mascoma and Connecticut Rivers to help workers produce clothing, furniture, tools, and many other household and agricultural staples in the 19th and early 20th centuries.

As Lebanon continued to grow, reaching a population of nearly 5,000 by 1900, so did its commerce. Hotels and taverns had served Lebanon's turnpikes since their earliest days and continued to host travelers and tourists as horse-drawn carriages gave way to gasoline-powered engines. Commercial centers in the central village and West Lebanon expanded with the introduction of new grocers, clothing retailers, and manufacturers of household goods. As the town's population increased, so did the number of students, and new, multistory schools were built throughout Lebanon. The growing town played host to increasingly popular events in the 20th century, with grand celebrations for the 150th anniversary of the town's charter and the conclusion of World War I.

The end of Lebanon's manufacturing era produced a dramatic downturn in the town's fortunes, but Lebanon's residents banded together to preserve the community. Lebanon became a city in 1958, and new industries rose in the place of the historic mills, utilizing state-of-art technologies that would define business in Lebanon for the rest of the 20th century. When a catastrophic fire demolished the central village's commercial district in 1964, the city embraced a modern vision for its reconstruction, and the district was rebuilt as a pedestrian mall as part of the nationwide urban renewal movement. When Interstate 89 opened in Lebanon in 1966, the city's long-established role as a regional crossroads was ensured for a new generation.

This book highlights Lebanon's many achievements and challenges, its notable personalities, and some of its most enduring and memorable institutions, arranged in eight thematic chapters. The first chapter focuses on the community's many beginnings—the impact of the glacial retreat 13,000 years ago, the area's first human residents, the Abenaki people, and the first European colonists. The second chapter explores the roads and railroads that would define Lebanon, from the earliest horse roads to the Westboro railyard to Interstate 89. The third chapter is dedicated to Lebanon's most influential and renowned individuals, from industrial innovators to remarkable artists. The fourth chapter is dedicated to the Mascoma River and Lebanon's many water-powered mills that defined the city for decades. The fifth and sixth chapters focus on the daily lives of the city's residents—"Lebanon at Work" highlights the city's many business and commercial enterprises, and "Lebanon at Play" focuses on the city's leisure activities and forms of recreation. Chapter seven features a selection of the city's most notable structures and landmarks, including residential, commercial, public, and private structures. Lastly, the eighth and final chapter covers the many fires that wracked Lebanon, which would ultimately shape and define the city.

Lebanon is not the first book about this city, nor is it the first to feature Lebanon's history in images. Nevertheless, it has been a great privilege to curate this brief glimpse into the history of Grafton County's most remarkable (and only) city, and to produce a new contribution to the very short list of publications entirely dedicated to Lebanon and its achievements—as well as its sometimes incendiary failures.

One

BEGINNINGS

Lebanon has had many beginnings, most of them lost to history. One of its first was around 11,000 BCE, when the Laurentide Ice Sheet, which covered millions of square miles across North America under a mile of ice, retreated north to once again expose Lebanon's soil to the air. Another beginning occurred around the same time, when humans first settled in what would later become New Hampshire, taking advantage of the newly accessible land and the influx of wildlife.

Another beginning was when glacial Lake Hitchcock, which covered much of West Lebanon, drained around 10,000 BCE. As told by the Abenaki, Ktsi Amiskw, the Great Beaver, built a dam to block the Connecticut River, hoarding the waters for himself and flooding the upper river valley, causing drought in the lower valley. The warrior Gluskabe fought and defeated Ktsi Amiskw, releasing the river to nurture life downstream and uncovering rich farmland upstream.

Untold other beginnings followed as the descendants of New Hampshire's original inhabitants grew and thrived. They developed and perfected farming along the Connecticut and Mascoma Rivers, managed the wildlife that inhabited their hunting grounds, and developed community identities, with the Koasek to the north and the Sokoki to the south, all members of the Western Abenaki. One of their last beginnings was their first contact with the Europeans, who arrived initially as traders and explorers, and later as invaders and colonizers.

The beginning about which we know the most was when European colonists began moving to Lebanon in 1761. The first of these settlers were already familiar with the area, having passed through Lebanon via the Connecticut River during the French and Indian War. The colonists petitioned Gov. Benning Wentworth for a charter for a new town in 1761. He granted the request despite the fact that the land was unceded and still belonged to the Abenaki people. Four colonists wintered in Lebanon in 1762, and the first colonial family moved to Lebanon the following year. By 1767, Lebanon's population was already 162, ushering in 20 years of growth: by 1790, Lebanon boasted a population of 1,180.

Humanity has called the Connecticut River Valley (pictured) home since as early as 11,000 BCE. Paleoindians moved into the area as the Laurentide Ice Sheet retreated at the end of the Pleistocene Ice Age. New Hampshire would be consistently inhabited for the next 13,000 years, predominantly by the Western Abenaki during the Woodland period, which lasted from 800 BCE until the arrival of European colonists around 1600 CE.

The confluence of the Mascoma (seen here near Gerrish Isle) and Connecticut Rivers made Lebanon valuable territory for the Western Abenaki—it was easily accessible by water and had rich soil, a relatively long growing season, and abundant wildlife for hunting. The Abenaki lived seasonally in Lebanon, returning to their winter villages for the long, harsh season.

Lebanon's Abenaki residents, similar to the couple in this 18th-century illustration, likely identified as members of a smaller Abenaki community such as the Koasek (centered around Newbury, Vermont) or the Sokoki (centered around Northfield, Massachusetts). These communities suffered massive population loss in the 17th and 18th centuries as European colonists brought catastrophic illnesses to New England. (Archives de la Ville de Montréal.)

Wigwams, like the ones in this mid-19th century illustration of an Ojibwe village, would have housed the Western Abenaki living seasonally in Lebanon. Easy to construct and dismantle, these structures and their sites of inhabitation left little trace as the Abenaki fled from the invading European colonists—which led to the colonists wrongfully declaring the lands as "uninhabited" upon their arrival.

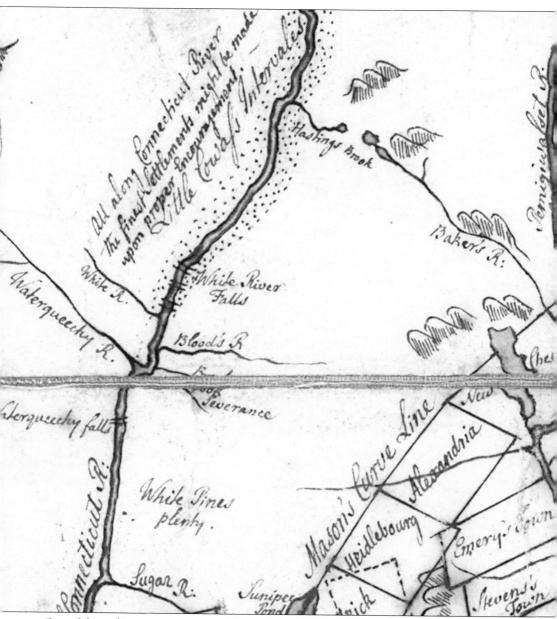

One of the earliest surviving maps of the Upper Valley is Joseph Blanchard and Samuel Langdon's 1756 "An Accurate Map of His Majesty's Province of New Hampshire." In this detail of the map, Lebanon can be identified by the "White River Falls" (modern Olcott Falls) in the northwest and "Blood's River" (modern Bloods Brook) to the south. The British charter officially demarcating Lebanon was issued five years after the creation of this map. The commentary on the Vermont side of the Connecticut River identifies this area as the "Little Cowass Intervales"—Koasek lands that European colonists had set their sights on for "the finest settlements." The presence of "white pines plenty" near Cornish and Plainfield was an additional draw to Europeans, who required strong pine trees for ship masts. The European colonists would ultimately "claim" the entire territory, expelling the Koasek and Western Abenaki who had lived there for millennia. (Library of Congress.)

Lebanon's first European colonists came from the Connecticut towns of Lebanon, Norwich, and Mansfield. Benning Wentworth, royal governor of New Hampshire, issued the charter for Lebanon and over a dozen nearby towns on July 4, 1761. The original proprietors, recipients of the governor's charter, divided the town into 100-acre lots and smaller allotments of fertile land along the rivers, as shown in this late-19th century version of the original map.

Lebanon's European settlers arrived via the Connecticut River but immediately constructed roads to connect with neighboring towns and, most importantly, Fort No. 4 in Charlestown. These early roads were known as horse roads, as they were appropriately sized for horses but too narrow for wagons. These roads followed the same routes as those used by the Western Abenaki, including the trails that followed the Connecticut and Mascoma Rivers.

The first years after Lebanon's charter was signed were spent clearing land and building roads. By 1763, families were starting to move into the township. Lebanon's earliest surviving residence was constructed on the shore of the Connecticut River around 1765 by Jonathan Dana. The Dana House stood on South Main Street in West Lebanon for over 200 years; in 1988, it was saved from demolition and relocated to Seminary Hill.

Lebanon's first meetinghouse, which served the civic and religious needs of the town, was built in 1772 at the top of Seminary Hill in West Lebanon. It was a small utilitarian building, similar to this early meetinghouse in Marlboro, Vermont. It served Lebanon—first in West Lebanon and later on Farnum Hill—until 1792. (Beyond My Ken, under CC BY-SA 4.0.)

The first meetinghouse served the early settlement well, but by 1790, Lebanon's population was nearly 1,200, and the small structure was no longer adequate. Furthermore, there was widespread disagreement regarding its location. Although the meetinghouse was originally constructed in West Lebanon, the site was not convenient for the town's residents in either the central village or East Lebanon, all of whom preferred a more easily accessible location. Around 1791, the discontented parties finally took action. Per Charles Downs, author of Lebanon's first history, the meetinghouse "was on the long-contested spot [on Seminary Hill] in early evening. It was not there in the morning." A well-organized mob dismantled the meetinghouse overnight, and it was soon rebuilt on Farnum Hill. The new location failed to resolve the dispute, however, and the matter was only settled when Robert Colburn offered the town a patch of land on the condition that it host the new meetinghouse. In 1793, the new meetinghouse (pictured) was built in the common, which would come to be known as Colburn Park.

The meetinghouse remained in Colburn Park until 1850, when it was moved slightly north to North Park Street, on the site of the current city hall. The 75-year-old structure was extensively renovated according to Victorian tastes in 1868, as shown here, and continued to serve as the center of Lebanon's government and culture until it was lost to fire in 1923.

As Lebanon grew, residents built schoolhouses throughout town to provide the area's children with easily accessible education. This school served the Hough district on the Meriden Road (now NH Route 120), and is typical of Lebanon's early schools. After it was retired from use as a schoolhouse, this building was converted into a private residence—a peaceful end similar to those of many of Lebanon's schools.

Two

ROADS AND RAILS

When the first Europeans came to Lebanon, they arrived via the Connecticut River. The Abenaki had established routes throughout the area, particularly along the Mascoma and Connecticut Rivers, and the first European colonists likely used these routes for their transportation network. The first roads built after Lebanon's chartering in 1761 were the horse road connecting Lebanon with Fort No. 4 in Charlestown and the King's Highway, which stretched from the horse road in West Lebanon east to the Enfield town line. These first roads, which were little more than narrow trails, were soon followed by dozens more, connecting homesteaders with their neighbors and the earliest mills. By 1810, the Fourth New Hampshire and Croydon Turnpikes had been completed, intersecting in front of the meetinghouse in what is now Colburn Park. Lebanon was established as a crossroads—a role it would retain as the railroad developed.

The Northern Railroad reached Lebanon in 1847, connecting Lebanon with Concord and Boston to the southeast and, several years later, with Burlington and Montreal to the northwest. Construction was a complicated matter, and workers blasted through long stretches of ledge and built numerous bridges over the Mascoma River, as well as one long bridge across the Connecticut River.

When the railroad officially opened on November 17, 1847, thousands greeted the first train to Lebanon. Among its passengers was Daniel Webster, who marked the day with a celebratory speech at the Lebanon passenger station. The railroad transformed the trip to Concord from a full day's journey into one that took several hours and reduced the trip to Boston from six days to just one. The effect on Lebanon's residents was immediate and dramatic, but the effect on its industry was incalculable. Lebanon's industry was no longer restricted to the local market or subject to costly and slow transportation by river or via an unreliable road system. Thanks to the railroad, Lebanon's mills and manufacturing (not to mention its population) boomed in both number and scale throughout the second half of the 19th century.

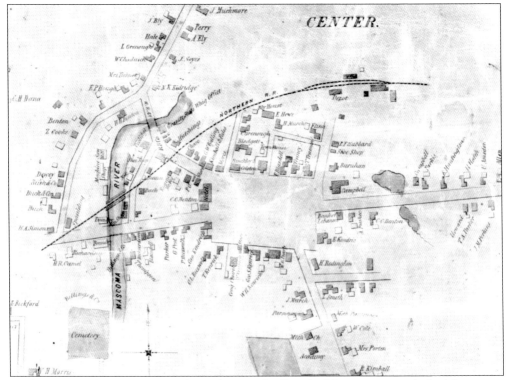

By the time of this 1855 map, Lebanon's central village had become a civic hub. After it was originally formed by the intersection of the Fourth New Hampshire (now US Route 4) and Croydon Turnpikes (now NH Route 120), a thriving village had grown around the crossroads. The turnpikes were rerouted around 1850, allowing the common (modern-day Colburn Park) to be fenced in as a dedicated park and community space.

This early photograph shows the intersection of the Fourth New Hampshire and Croydon Turnpikes before they were rerouted around the park. The open land in the left foreground was converted to Colburn Park, while the buildings in the background, including the Lebanon Congregational Church (at left), would eventually front South Park and West Park Streets.

One of Lebanon's first roads was the King's Highway—one of many roads in New England to be given this moniker. Lebanon's King's Highway ran from the horse road by the Connecticut River east over Farnum and Storrs Hills and continued to the Enfield town line. This c. 1875 photograph shows the road still in use as it cuts through the orchards on top of Farnum Hill.

The earliest road to East Lebanon followed the path of the modern Route 4 over Fellows Hill, but by the middle of the 19th century, there was a clear need for a lower road following the Mascoma River. The Mill Road fulfilled this need and became a popular scenic route, as shown in this 1929 photograph by George U.L. Leavitt.

The first bridge across the Connecticut River between West Lebanon and Hartford, Vermont, was built by Elias Lyman (1768–1830) around 1803. The crossing was a toll bridge, as was the three-span covered bridge that replaced it in 1836, which is shown here. The town of Lebanon purchased the bridge in 1879 and eliminated the tolls. The covered bridge stood until 1895.

The second covered bridge built by Elias Lyman was replaced by this iron bridge in 1897. The new three-span Platt truss bridge cost nearly $25,000, which included the removal of the existing bridge and a temporary span utilized while the iron bridge was being constructed. The replacement bridge stood for 39 years until it was severely damaged by the 1936 flood. Over 700 bridges in New England were damaged or destroyed by that flood.

The West Lebanon railyard—later known as Westboro—was connected to White River Junction, Vermont, and its rail network by a roughly 500-foot rail bridge across the Connecticut River. The original 1848 bridge, designed by Henry R. Campbell, was replaced in 1871 by this wood lattice and arch bridge, which stood until 1892.

In 1892, a continuous deck plate girder bridge replaced the 1871 bridge across the Connecticut River. This replacement bridge was damaged by flooding during the November 1927 hurricane. Filled coal cars were parked across the bridge (shown here) to stabilize it against the debris carried downstream and subsequently caught by the structure. The bridge survived, but two men died while working to clear the debris.

Once the railroad came to West Lebanon in 1848, it became the village's defining feature. Residents worked aboard the locomotives, built and maintained the railyard, and labored in the booming agricultural and manufacturing industries that were connected via the railroad to Boston, Burlington, and beyond. This 1866 view of West Lebanon from across the Connecticut River in White River

Junction, Vermont, shows the village as it embraced its role as a railroad hub. The original railroad roundhouse is to the right of the bridge on the far riverbank, with the village stretching up the hillside beyond it. The Tilden Ladies' Seminary (later the Rockland Military Academy, then West Lebanon High School, and finally Seminary Hill School) is visible at upper right.

When the Northern Railroad came to Lebanon in 1847, it provided the town with rapid and affordable access to the major cities of the northeast. The Lebanon passenger depot, shown here in 1927, was located at the end of Campbell Street and served the central Lebanon village for over a century before it was torn down in 1965.

The East Lebanon (later known as Mascoma) station served the easternmost part of Lebanon, once an important industrial and population hub. The passenger depot stood on the eastern shore of Mascoma Lake. It was accompanied by a freight depot and an icehouse, which was used to store ice cut from the lake before it was transported via train to Boston.

The West Lebanon railroad station was originally and aptly called West Lebanon. After a fatal accident in 1908 in Haverhill, New Hampshire, in which the operator confused Haverhill and East Haverhill, the West Lebanon station was renamed Westboro, and the East Lebanon station's name was changed to Mascoma. This photograph, taken sometime after 1908, shows the Westboro passenger depot, with the back of the structure at 35 South Main Street in the background.

The Westboro rail yard, shown here around the mid-20th century, marked the northern terminus of the Northern Railroad and was fully equipped to service the line's engines. This roundhouse, built in 1929, was the third constructed at Westboro and had 24 stalls arranged around a 110-foot turntable. The roundhouse fell into disrepair in the final decades of the 20th century and was demolished in 2021.

This photograph, with a view facing southwest, was taken sometime before 1887 from the top of the Hanover Street dry bridge (now the pedestrian mall). After leaving the Lebanon passenger depot, the Northern Railroad proceeded west behind the buildings fronting North Park Street and curved around the north end of Court Street to pass under Hanover Street. Workers are shown repairing the rails at the Mill Street crossing, visible just in front of the covered bridge in the distance. This covered bridge spanned the Mascoma River, and the rail line continued across the bridge and into the intersection with High Street and Mascoma Street (later known as Jones Crossing). The railroad siding to the left served the Mead, Mason and Co., and Kendrick and Davis factories on Water Street. Nearly all of the structures in this image were destroyed in the 1887 fire; only the houses in the extreme distance beyond the covered bridge would survive Lebanon's first great conflagration.

This 1949 picture shows the same view as the previous image but more than 60 years later. Note the offices of the *Granite State Free Press* on the right and the deck rail bridge across the Mascoma that replaced the covered bridge. As with the structures shown in the previous image, most of these buildings were later destroyed by fire, but this time in 1964.

After the 1964 fire, which devastated Lebanon's downtown, urban renewal and modernization led to a total redesign of the area. This photograph was taken in 1969, from near the same location as the previous two images, in the midst of the construction of a railroad tunnel under the future pedestrian mall and State Route 120. Mill Street was later entirely removed, and the commercial buildings were incorporated into the pedestrian mall.

The stone arch underpass, built in 1848, is one of three Lebanon structures listed in the National Register of Historic Places. The underpass carried the Northern Railroad from Lebanon to West Lebanon while allowing traffic to pass on Glen Road below it. Northern Railroad engineer Henry R. Campbell, for whom the Carter House was constructed, is often credited with designing the bridge.

West Lebanon, one of Lebanon's first commercial districts, was ideally situated on the banks of the Connecticut River and along the Fourth New Hampshire Turnpike after it crossed from Vermont into New Hampshire. This 1915 view looking south shows some of West Lebanon's commerce as well as its wide, tree-lined avenue, which was largely lost in the late 20th century.

This iron "highway" bridge once stretched across the Connecticut River and connected Wilder, Vermont, with the East Wilder neighborhood of Lebanon. The Wilder Paper Mill, constructed in 1884, used this bridge to reach Lebanon and its pulp-processing factories for decades before the bridge was dismantled in the 1940s as part of the construction of the Wilder Dam downstream.

This unusual arrangement of bridges occurred on Route 4 near Glen Road. The Hubbard road bridge (left) crossed the Mascoma River. The covered rail bridge (right) was called the Low Bridge, in contrast with the road bridge (visible through the covered bridge) that ran over the rails and was known as the High Bridge. The two road bridges were consolidated in 1921 and now form the Terri Dudley bridge.

The Northern Railroad passed over the Mascoma River 14 times between East Lebanon and West Lebanon. Many of the bridges were covered, including the bridge on the right, which was at the intersection of Bank Street and Pumping Station Road. This train is crossing Bank Street just west of Benton Bridge (left) on its way toward the Lebanon central village station, which was located less than one mile west of this crossing.

On April 30, 1886, the *Atlantic* attempted to cross the bridge at Chandler's Mill (near Riverside Drive). However, the bridge was designed for older, lighter trains, and the 60-ton locomotive caused the bridge to buckle. The engine crossed before the bridge failed, but three freight cars were caught in the collapse. There were no serious injuries, and, remarkably, the wreck was cleared and a temporary bridge was installed by the following evening.

Just east of the Chandler's Mill bridge, the Northern Railroad (right foreground) passed beside the Packard covered bridge near Baker's Crossing (on Riverside Drive). Until the early 1930s, a road connected this area to LaPlante Road, which connected to the Croydon Turnpike (Route 120). The overpass carrying this road over the railroad is visible in the middle distance at right.

Of the many Mascoma River crossings that the railroad made between East and West Lebanon, two of the most distinctive were the twin bridges at Mill Road, shown here from the northeast. For a short distance (demarcated by a covered bridge at each end), the rail line ran alongside Mill Road, identified in this c. 1880 photograph by the horse (with sleigh) drinking from a watering trough.

Lebanon's town sheds, supporting the department of public works, were located on Spencer Street for much of the 20th century. This photograph of four town workers and their trucks was featured in the 1945 Lebanon Town Report. The garages shown here were destroyed by fire only five years later, but the rebuilt facility continued to serve Lebanon and its roads until 2008.

Appropriately, Hanover Street originally connected Lebanon's central village with the neighboring town of Hanover to the north. The stretch of Hanover Street from Colburn Park to Hough Square developed as one of Lebanon's main streets, teeming with shops and local businesses. This 1961 image shows the north side of the street as it existed before the 1964 fire, featuring Tom's Toggery, the Lamplighter Restaurant, Woolworth's, and McNeill's Drug Store.

After the 1964 fire destroyed the northwest half of the Hanover Street shopping district, it was declared a disaster area, and federal urban renewal funding supported the street being converted to a pedestrian mall. This picture from September 1969 shows the closed street from Colburn Park, with the roadway partially torn up and the installation of decorative planters underway.

The pedestrian mall was completed by the following summer, and hundreds of people attended the opening ceremony on August 4, 1970. The previously busy main street was converted to a walking retail district—one of nearly 200 pedestrianization projects across the country. The project was highly controversial at the time and for decades afterward, but nationally, Lebanon's mall is considered a successful implementation of the pedestrian model.

Construction of Interstate 89 significantly altered the landscape around Lebanon—though not through the central village, as one plan had proposed. This May 1966 photograph shows the highway construction between exits 18 and 19 and looks northeast from the newly built Mascoma Street bridge. The southbound rest area was built to the right of the most distant point of the roadbed.

Three

NOTABLE LEBANONIANS

As one of the smallest cities in one of the country's smallest states, Lebanon should have very little claim on famous people of note. For its size, however, Lebanon punches above its weight and is privileged to call many diverse and fascinating individuals Lebanonians. From local community leaders like Phineas Parkhurst and Frank Churchill to figures of national fame like Joseph Smith and Phineas Gage to artistic and cultural innovators like Ammi Burnham Young and Alanis Obomsawin, Lebanon has been home to many individuals who have left a lasting impact.

This chapter highlights a few such figures, offering a brief glimpse into their lives and the impact they had on their community, whether local, regional, or national. Selected for their unique accomplishments, unusual experiences, and exceptional contributions to society, these notable Lebanonians emphasize the importance of Lebanon in the region and its role in the development of the Connecticut River Valley region, New Hampshire, and the United States across more than two centuries.

Lebanon's version of Paul Revere was Phineas Parkhurst (1760–1844). In 1780, Parkhurst was shot during a conflict between European colonists and Western Abenaki in Royalton, Vermont. Parkhurst rode 25 miles to warn Lebanon's residents of the danger. Lebanon doctor Ziba Hall treated Parkhurst's injuries, and Parkhurst ultimately remained in Lebanon, studying medicine under Hall and becoming Lebanon's preeminent physician and a leading civic figure.

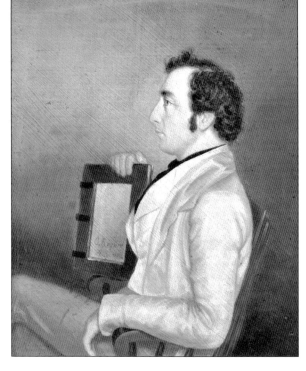

Ammi Burnham Young (1798–1874) was the first supervising architect of the US Treasury Department. He was born in Lebanon and designed a number of local buildings, including the Lebanon Congregational Church and buildings at Dartmouth College and the Enfield Shaker village. As supervising architect, Young designed dozens of customhouses, post offices, and other federal buildings for cities across the country. (Vermont Historical Society.)

Joseph Smith (1805–1844) was born in Sharon, Vermont, and moved to Lebanon around 1812. While in Lebanon, he contracted typhoid fever, which infected a bone in his leg. He was fortunate to be treated by Dartmouth Medical School founder Nathan Smith, who excised the infected area and saved Joseph's leg from amputation. Joseph Smith went on to found Mormonism and the Latter-Day Saints movement, and he became the religion's first prophet.

A noted artist of the White Mountains, Benjamin Champney (1817–1907) moved to Lebanon in 1827 to live with his aunt and uncle. He worked in a cotton factory alongside another local boy, William Bliss, for four years before going on to study art. Champney was renowned for his paintings of the White Mountains, and he founded both the North Conway art colony and the Boston Art Club.

William Bliss (1815–1853; standing at left) moved to Lebanon at the age of 7 and enrolled at West Point at 14. He was chief of staff to Maj. Gen. Zachary Taylor (seated) and became Taylor's private secretary after the general became president of the United States in 1849. (National Portrait Gallery, Smithsonian Institution; partial gift of the Quaker Oats Company.)

William Bliss was a talented linguist able to read 13 languages, and he taught mathematics at West Point. He fought in the Mexican-American War and participated in the forced displacement of the Cherokee people. Bliss died of yellow fever at the age of 37, and the US Army named the El Paso military post Fort Bliss—where this monument to Bliss stands—in his honor. (Kevin White, www.hmdb.org.)

Phineas Gage (1823–1860) was born in either Lebanon or neighboring Enfield. He gained international fame after surviving a blasting accident that launched a tamping iron (shown here) through his head. Gage went on to become an important neuroscientific subject. (Warren Anatomical Museum collection, Center for the History of Medicine in the Francis A. Countway Library of Medicine, Harvard University. Gift of Jack and Beverly Wilgus.)

In 1848, Phineas Gage was conducting blasting for the railroad near Cavendish, Vermont, at a site much like this contemporaneous cut in Lebanon. A detonation misfire propelled his tamping iron through his head, landing approximately 80 feet away. Gage was able to sit up and speak within several minutes, and upon encountering a doctor 30 minutes later, declared, "Doctor, here is business enough for you."

George Storrs (1796–1879) was a preacher and antislavery activist. Storrs was born in Lebanon, and he lectured on the evils of slavery throughout New Hampshire and Vermont. In 1835, he was arrested in the middle of a lecture, and a mob threatened to lynch him for his abolitionist preaching. Storrs persevered, however, and went on to become an influential Christian writer and an Adventist leader.

Ticknor and Fields, founded by Lebanon native William Davis Ticknor (1810–1864), was one of the most important American publishers of the 19th century. The company was responsible for publishing the *Atlantic Monthly* and authors like Alfred Tennyson, Charles Dickens, and Nathaniel Hawthorne. Hawthorne and Ticknor would become close friends. When Ticknor developed pneumonia and died in 1864, Hawthorne was severely affected, and the author passed away about one month later.

Henry Wood Carter (1822–1897) started his remarkable career as a traveling salesman, selling household goods from his ornate wagon. He ultimately began producing his own goods, particularly overalls and other work clothing. The H.W. Carter & Sons factory dominated the Lebanon manufacturing scene for more than a century. The Carter House and the former Carter factory continue to define Bank Street in the 21st century.

Thomas Penick (1830–1906) and Mona White Penick (1830–1915) were born enslaved in Kentucky. They freed themselves, along with their daughter Harriett (1858–1918), in 1865, and Thomas enlisted in the Union army in the Civil War. After the war, the Penicks moved to East Lebanon, living in a house overlooking Mascoma Lake, and Thomas worked as a section hand for the railroad until his retirement in 1896.

Lebanon native John Milton Thompson (1842–1922) enlisted in the Union army at the start of the Civil War and, in 1862, helped lead the First South Carolina Infantry Regiment—one of the first black regiments, which was composed of formerly enslaved men. He retired from the army more than 40 years later as a brigadier general, having served in the Spanish-American and Philippine-American Wars.

Many of the most important 19th-century photographs of Lebanon were taken by William W. Culver (1834–1927), shown here around 1910 in his later career as a housepainter. Culver studied art in Boston under Alexander Ransom and learned photography in Alabama during the Civil War. Culver was purportedly responsible for the only photograph of Jefferson Davis's inauguration as president of the Confederate States of America.

"Of course every body [sic] knows Charlie—that same Old Charlie who has driven all over the roads in California," reported J. Ross Browne in his *Harper's Weekly* story "Washoe Revisited" in 1865. Browne was describing Lebanon native Charley Parkhurst (1812–1879), a regional celebrity in his role as a stagecoach driver—much like the unidentified New Hampshire stagecoach driver shown here—during the California Gold Rush. Parkhurst was assigned female at birth (and known as "Charlotte" in Lebanon) but lived most of his life as a renowned and fearless stagecoach driver on some of the country's most dangerous roads. After his death, Parkhurst's physical attributes became public, and national publications, including the *New York Times*, breathlessly reported on the "female" stagecoach driver. In his article, Browne marveled: "But I had implicit confidence in Old Charlie. The way he handled the reins and peered through the clouds of dust and volumes of darkness . . . when I could scarcely see my own hand before me, was a miracle of stage-driving." (Keene Public Library.)

This group of prominent Lebanon businessmen and civic figures was photographed sometime around 1890 in front of the Lebanon town hall. From left to right, they are Lyman Whipple, Owen Briggs, James Ticknor, John S. Freeman, William P. Burton, Charles Downs, Frank Churchill, and William F. Shaw.

Walter Arlington Latham (1860–1952), better known as Arlie Latham, played professional baseball from 1880 until 1909. Born in West Lebanon, Latham is believed to be the first Major League Baseball player from New Hampshire. He holds the seventh all-time record for most stolen bases, totaling 742 over his career. After retiring, Latham became a coach and organized baseball in England, and purportedly taught King George V how to throw and catch.

Between 1879 and 1881, Lebanon High School students had the privilege of studying biology with Nettie Stevens (1861–1912), an early geneticist responsible for discovering sex chromosomes. Nettie taught in Lebanon before pursuing higher education, culminating in a PhD in cytology from Bryn Mawr College. Her research identified X and Y chromosomes and normalized the use of fruit flies in genetic research. (Special Collections, Bryn Mawr College Libraries.)

George Halsey Perley (1857–1938) was born in Lebanon and moved to Ottawa, Ontario, as a child. Perley was vice president of the Canada Atlantic Railway Company and was elected to the Canadian Parliament in 1904. During World War I, he served as minister of the Overseas Military Forces and was very briefly Canada's secretary of state in 1926. (Library and Archives Canada/Department of External Affairs fonds/c052327.)

Frank Carroll Churchill (1850–1912) boasted a resumé to rival the best of his peers. His remarkable career started when he left H.W. Carter to cofound Carter & Churchill in 1877, producing overalls and outerwear. He went on to serve in the New Hampshire House of Representatives and was chair of the state delegation to the National Convention in 1888, helping to nominate Benjamin Harrison for president. In 1899, Churchill was responsible for securing the charter for the Mascoma Savings Bank; he then served as the bank's president. He rose to national prominence as an employee of the US Department of the Interior, where he was appointed revenue inspector to the Cherokee Nation in 1899, which sent him and his wife, Clara, to Oklahoma. He was shortly thereafter appointed as a special agent and visited more than 80 Native American communities. His final assignment was to Alaska, following his appointment by Pres. Theodore Roosevelt as a special agent of education. Poor health forced Churchill to resign in 1909, and he died in Lebanon in 1912.

Clara Corser Turner Churchill (1851–1945), Frank Churchill's wife, boasted interests as varied as her husband's. Born in Boscawen, Clara Churchill moved to Lebanon at the age of 11 after the death of her parents. One of her many interests was music; she was the organist of the Lebanon Unitarian church for more than 25 years and also composed music. Churchill was an essayist and wrote about the 15 states she and her husband visited during his government work. Her most significant interest, however, was in the artistic and cultural objects produced by the more than 80 Native American and Indigenous communities that the Churchills encountered. Churchill collected more than 1,400 objects, including woven baskets, carved spoons, fishing tools, jewelry, and sculptures, from communities including the Iñupiak, Tlingit, Navajo, and Haudenosaunee. Churchill had hoped that her house and collection could be turned into a museum, but there were insufficient funds. After her death in 1945, Churchill's remarkable collection was donated to the Hood Museum of Art at Dartmouth College, where the items became the majority of the museum's Native American and Indigenous collection.

One of New Hampshire's most important suffragists was Mary Inez Stevens Wood (1866-1945). Wood lived in Lebanon for several years after her marriage and went on to help lead many state women's organizations. When the 19th Amendment (to give women the right to vote) came before the New Hampshire Senate in 1919, Wood was the only speaker allowed to testify in support of the amendment. After her testimony, the amendment was ratified.

Norris Cotton (1900-1989) was a Lebanon lawyer who became one of the city's most successful politicians. He represented New Hampshire in the US House of Representatives and the US Senate from 1947 to 1975. Cotton helped found the Cancer Center at the Mary Hitchcock Memorial Hospital, which bore his name for nearly half a century. (Library of Congress.)

Lane Dwinell (1906-1997), New Hampshire governor from 1955 to 1959, established himself in Lebanon business and government before successfully dominating state politics. Dwinell served in the state house of representatives and as state senate president before becoming governor. He went on to serve as assistant secretary of state for administration under Pres. Dwight Eisenhower, and he held the dubious honor of officially announcing Richard Nixon's reelection campaign. (National Archives.)

Patrizia Cobb Chapman (1927-2010), better known as Buff Cobb, was an actress, Broadway producer, and cohost of *Mike and Buff*, one of television's first talk shows, which Chapman hosted with her then-husband Mike Wallace. As a producer, she was nominated for a Tony Award for her revival of George Bernard Shaw's *Too True to Be Good*. Chapman retired to Lebanon, where she died in 2010.

After his distinguished career as director of the National Institutes of Health Clinical Center and president and dean of the Mount Sinai School of Medicine, Thomas Clark Chalmers (1917–1995) moved to Lebanon. Chalmers was best known for developing and promoting the randomized controlled trial in medicine. In retirement, he served as chairman of the board of directors of the Dartmouth Hitchcock Medical Center. (National Library of Medicine.)

Highly acclaimed filmmaker Alanis Obomsawin (born in 1932) has made more than 50 documentary films about Indigenous and First Nations peoples. Obomsawin was born in Lebanon but moved to the Odanak Abenaki First Nations reserve in Quebec, Canada, before her first birthday. She has received dozens of awards and was appointed Companion of the Order of Canada, the country's highest civilian honor. (Kat Baulu, under CC BY 2.0.)

Four

THE MASCOMA
AND THE MILLS

Lebanon's first mill was a sawmill built by Oliver Davison in 1763, harnessing the power of the Mascoma River to fulfill the fledgling village's need for sawn wood. It would take three years before the next mill, a gristmill, was constructed. Until the completion of Lebanon's first gristmill, residents would transport their grain to Fort No. 4 in Charlestown, New Hampshire—a round-trip of 70 miles, testifying to the vital importance of a gristmill to the fledgling community.

These early mills were soon joined by other sawmills and gristmills, as well as by cloth mills. By 1817, Lebanon was home to 20 mills relying on the energy of the Connecticut and Mascoma Rivers to power their production. As the 19th century progressed, industrial centers developed across Lebanon: in East Lebanon at the foot of Mascoma Lake, at the foot of Benton Hill west of Colburn Park, in Scytheville at the base of Slayton Hill, and in Butmanville near the confluence of the Mascoma and the Connecticut. These mills produced furniture, metal tools, machinery, and—increasingly, as the century drew to a close—woolen textiles.

Lebanon's diverse industries were greatly affected by fire during the 19th century. The first large-scale fire devastated East Lebanon in 1840, destroying the burgeoning hub and permanently relegating the area to the shadow of the central and West Lebanon villages. The second such fire in 1887 leveled the central village and nearly all of its industry. However, this fire proved insufficient to quash the village's industry, which was largely rebuilt within five years. It did, however, serve to concentrate manufacturing into the woolen mills that would dominate Lebanon's industry for the next 75 years.

This detail from George Norris's 1884 hand-drawn bird's-eye view of Lebanon shows the concentration of mills and industry along the Mascoma River in the central village. Nearly all of the available land along the river was occupied by mills, manufacturing, or other industries. The Hanover Street bridge appears at the top center, and factories and manufacturing structures line the river south to the covered rail bridge and the Mascoma Street bridge at center—these include Rix Woodworking (labeled 24), Mead, Mason and Co. (14), and S. Cole & Son Foundry (19), among many others. Nearly all of the buildings along the river were destroyed only three years later during Lebanon's catastrophic 1887 fire. Colburn Park is visible at top right, with the town hall (6) and newly constructed Whipple Block (30) facing it. (Library of Congress.)

The first gristmill in Lebanon was built by John Bennett near the confluence of the Mascoma and Connecticut Rivers around 1765 on the site of the modern Powerhouse Mall. The mill operated for over 150 years in various forms, adapting to meet the needs of the developing town. Thomas Waterman ran the site as a sawmill in the late 19th century, as shown here.

The area around the intersection of Slayton Hill Road and Mechanic Street was once an industrial hub known as Scytheville. Two of the neighborhood's primary manufacturers were the Mascoma Edge Tool Company, which manufactured scythes (shown here at the rear left), and the G.W. and M.L. Stearns Manufactory, which produced scythes and other tools (occupying the buildings in the foreground).

The dam at the foot of Mascoma Lake (right foreground) provided power for a succession of mills, starting with Elisha Payne's sawmill in 1780. Cotton and woolen mills followed in the 1830s, and the W.O. Haskell & Son furniture factory was established in the wake of the 1840 fire that devastated East Lebanon. This photograph was taken after the Haskell factory closed in 1876, with the abandoned shop shown at left.

Frank Kendrick and William Davis started Kendrick and Davis in 1876, soon moving into a factory at 12 Water Street. Here, they produced their signature "dust-proof" watch keys, which featured slots in the key tube to prevent dust and lint buildup. By the time of this photograph in 1885, the *Granite State Free Press* reported that the factory employed 85 workers, around 24 of whom are shown here.

The Mascoma River powered J.C. Sturtevant and Company's lower shops (right) and the Lebanon Woolen Mill (left). Sturtevant and Company manufactured furniture in this Water Street factory as well as in their upper shops on High Street until 1876, when the business was sold to Mead, Mason and Company. The Mascoma Street bridge at the foot of Benton Hill is visible in the distance.

After the 1887 fire destroyed the Mead, Mason and Company's Water Street factory, the site was sold to the Riverside Woolen Company. Construction of the Riverside Mill began in 1893. The new mill was built of brick and supported on robust stone foundations alongside the Mascoma River. The building survived both the end of Lebanon's mill era and the 1964 fire, and it continues to stand in the 21st century.

The Mascoma Mill, shown here in the foreground, was first constructed as a flannel mill in 1882 and was expanded as the American Woolen Mill in 1899. The mill's location near Scytheville preserved it from destruction during the 1887 fire that consumed Lebanon's other major mills. This photograph was taken from the side of Storrs Hill around 1915 and shows the mill along

the Mascoma River, with Mechanic Street running beside it and leading into the central Lebanon village in the distance. The towers of the town hall and the First Congregational Church are visible in the background at right, as is the smokestack of the Lebanon Woolen Mill (between the towers). The railroad runs just behind the houses lining Mechanic Street at left.

The Lebanon Woolen Mill was built around 1888 from the ruins of the Carter and Rogers mill, which was destroyed in the 1887 fire. The mill operated for nearly 75 years, manufacturing textiles until the collapse of the wool industry in the 1960s. The mill complex then housed the Kleen Laundry business for several decades until it closed in 2019.

At the Lebanon Woolen Mill, fabric was finished in the basement, weaving was performed on the first floor, and carding and spinning were executed on the second floor. This image shows the first-floor weaving room, with the looms connected to the overhead drive shaft via belts. The whole operation was powered by the Mascoma River, which ran to the east of the mill.

Lebanon's mills employed children as well as adults, which was common in the time before labor laws were passed in the United States. This photograph was taken in the "drawing in" room in one of Lebanon's woolen mills around 1890. Drawing in involved turning individual strands of yarn into the thicker fibers that would form the warp and weft of fabric—a critical step before the yarn could be put on a loom and woven into cloth.

Everett Knitting Works was built in 1891 between High Street and the Mascoma River. At its height, it employed approximately 300 people and produced undergarments for the federal government and for Sears, Roebuck & Company. The mill closed in 1931—a victim of the decline of the New England textile industry and the Great Depression.

This c. 1911 picture was taken facing north from the rail bridge over the Mascoma River. The upper dam shown here powered both Everett Knitting Works (left foreground) and the Flanders sawmill (left background). After Everett Knitting Works closed in 1931, the building became a tannery of the E. Cummings Leather Company, and it was ultimately demolished in 1981.

The Emerson (formerly Mascoma) Edge Tool Company began in Scytheville in 1856 before relocating to the Lebanon Slate Mill in East Lebanon in 1880. The Emerson Edge Tool Company manufactured scythes and other cutting devices on the bank of the Mascoma River near the modern intersection of Routes 4 and 4A until 1907, when the mill was sold and became a bobbin mill.

Like many factories, the Emerson Edge Tool Company kept cats to help protect the products from rodents and to limit the spread of animal-borne diseases. Unlike other factories, however, the Emerson Edge Tool Company documented their furry employee. This photograph of Dick, the mill cat, identified him as one of their valued employees.

The first paper mill in Wilder, Vermont, was constructed in 1865. After it washed away in 1872, the Olcott Falls Company took ownership and constructed a new paper mill (pictured) in 1884. A pulp mill was built across the river in East Wilder (a neighborhood of West Lebanon), providing the raw materials for the paper factory. The mills were connected via two bridges across the Connecticut River.

The paper mill had been closed for nearly 25 years by the time the Wilder Dam was completed in 1950. The new dam replaced an old power plant on the former site of the paper mill, which is just visible in this photograph (on the peninsula at center left). The dam cost $16 million and, at the time, was the largest project in Lebanon's history.

Construction of the Wilder Dam took two years and employed over 500 workers. This 1949 photograph shows the scale of the project, which spanned 2,100 feet between Wilder, Vermont, (opposite shore) and Lebanon and stood 59 feet high. The dam would raise the upstream water level by 15 feet, flooding 335 acres on the shore.

Five

LEBANON AT WORK

The mills were Lebanon's major employers for much of its history but represented only one avenue of work for Lebanon residents. In addition to sawmills and gristmills, many of Lebanon's earliest businesses were taverns and inns that provided room and board to travelers passing through town. Commercial shops soon joined these earliest businesses, and by 1800, hubs of commerce had developed in East Lebanon, the central village, and West Lebanon.

As the 19th century progressed, other forms of manufacturing developed away from the waterpower of the Mascoma and Connecticut Rivers. Shops and commercial spaces expanded to sell the results of the increasing production as well as to peddle goods from further afield. Many of Lebanon's stores were family-run and passed from one generation to the next for over a century. By the middle of the 20th century, however, national chains and nonlocal businesses began to compete for customers. At the same time, new forms of manufacturing arrived in the form of aerospace engineering and arc-plasma cutting, replacing the dying textile mills that had dominated industry in Lebanon for decades.

This chapter highlights some of these commercial and industrial developments as well as the owners and employees who made them the successful endeavors that they were—whether for 5 years or for 100.

East Lebanon was once Lebanon's primary industrial hub—a thriving mill and commercial center at the foot of Mascoma Lake. However, an 1840 fire decimated the community, and by the late 19th century, East Lebanon was little more than a small village with a rail depot on the edge of the lake, as shown here from the slopes of Mount Tug.

Sargent's Hotel stood on the corner of Bridge and Main Streets in West Lebanon. The hotel's origins stretched back to the 18th century, when William Dana first built a hotel on the site. Later known as the Allen House, the hotel served West Lebanon into the 1930s, when it was replaced by a service station.

The hotel at 14 Parkhurst Street, originally facing Depot Square across from the Lebanon railroad passenger station, was built by Frank Sayre (1841–1907) in 1877 and appropriately named Sayre's Hotel. Over the years, it regularly changed hands, becoming the Williamson House (shown here) in 1890, the Barnes Hotel in 1912, and the Lebanon Inn from 1929 until the hotel closed in 1945.

The brickyard at 174 Hanover Street operated from the late 18th century until 1974. For most of those years, it belonged to the Densmore Brick Company, shown here in 1886. Nearly every brick building in Lebanon, as well as many in neighboring towns, was constructed of Densmore brick, including the Soldiers Memorial, the Sacred Heart church, and many residences. In 1888, the brickyard produced 14,000 bricks per day.

The store owned by John K. Butman (1826–1905) is shown at left in this late-19th-century photograph and was the quintessential general store, offering groceries, building materials, clothing, and manufactured goods for sale. Located at approximately 120 South Main Street in West Lebanon, the store and its owners (who lived next door) would so define the area that the southwestern part of town was known as Butmanville for decades.

The quarry on Mount Support Road supplied pink granite for industrial and construction projects throughout New Hampshire and Vermont. In 1895, the year before this picture was taken, the *Vermont Journal* reported that "handsome specimens of granite" were being quarried for the footings of a new West Lebanon iron bridge across the Connecticut River and for a Passumpsic railroad bridge in Barnet, Vermont.

When Henry W. Carter (1822–1897) established his business on Bank Street in 1859, he sold his wares—everything from cigars to razors to perfume—from an elaborately painted horse-drawn wagon. Customers placed orders based on samples, which he filled from his warehouse (above right). Around 1870, Carter started selling overalls produced by Converse Cole in East Plainfield. When Carter began to sell more than Cole could produce, Carter bought the company and took over the manufacturing. His success led to the factory at 11 Bank Street, which was built as a warehouse in 1884 and expanded into a nearly 30,000-square-foot structure (below) in 1893. In 1904, H.W. Carter & Sons boasted more than 170 employees, and it remained one of Lebanon's major employers until it closed in 1985. The building would later house the AVA Gallery and Art Center.

Henry Carter's nephew, William Carter, joined the family business in 1865. By the end of the decade, however, he had started his own business, and in 1877, he joined forces with Frank Churchill to create Carter & Churchill. They produced overalls through World War I, as shown in this 1914 photograph, before focusing on outerwear. Relabeled Profile Skiwear in 1970, the business was a Lebanon mainstay until it closed in 1983.

George Edson (1850–1921) and Sherman Chadwick (1851–1894) were already the successful proprietors of Edson & Chadwick in White River Junction, Vermont, when they opened their shop on Main Street in West Lebanon in 1891. Initially focused on groceries and general provisions, the shop eventually shifted to baked goods and relocated to the foot of Seminary Hill, operating as the Corner Food Shop before leaving the Edson family in 1962.

Originally known as Packard Hill Farm, the farm at the top of the hill overlooking the bridge of the same name was long home to the Winona Dairy—first under the Pringle family and, starting in 1965, under Volney and Barbara (Peck) Slack. After the dairy closed, the property was sold to developers in the 1980s. Only the original farmhouse survives, nestled amongst the modern houses.

Snow-covered roads in the late 19th century were made passable not by plows but by snow rollers, which packed the snow down to allow horse-drawn sleighs to pass over the top. This snow roller team worked on North Park Street around 1896. The new Bank Block (constructed in 1893) is visible at far right.

Frank Adams (1885–1933) was born in Barnard, Vermont. After apprenticing there for blacksmith Charles Aikens, Adams moved to Lebanon and opened his own blacksmith shop at the corner of High and West Streets. This photograph was taken around 1905 and portrays Adams and his workers showing off their craft.

Members of the United Garment Workers of America labor union are shown here posing with their Labor Day parade float on North Park Street around 1905. Many of the union members were employees of the H.W. Carter & Sons factory on Bank Street, including Mary Josephine Stas (third from left), Eliza Copp (fifth from left), and Daisy Dowse (seated immediately to the left of the "U.G.W. of A. Local 90" sign).

Hunt's Department Store occupied the Baldwin Block at 20 Hanover Street for nearly 70 years. This c. 1910 photograph shows the store clerks at work, including Beatrice Painchaud, the store's head clerk (second from left), Ida Leavitt (at rear), and Julia Nourie (right). The store closed three years before the building was destroyed in the 1964 fire.

The first sawmill between High Street and the Mascoma River at the upper dam was built by Joel Amsden around 1819. The sawmill supplied Lebanon with lumber for the rest of the century, passing to Thomas Marston in 1882 and to the Flanders family in 1900. The sawmill is shown here around 1910. The sawmill finally shut down around 1921, when George Flanders opened the Flanders and Patch Garage.

Carrie Lowe (1856–1935) was an employee of Houghton & Macy's Ladies' Pavilion and Housekeeper's Emporium in the 1880s before taking over the business in 1890. The Carrie Lowe shop, located on North Park Street, sold ready-to-wear clothing and advertised itself as "a mecca for women." After the Converse Pavilion (pictured) was demolished in 1911, Lowe moved her store into the Hotel Rogers, where it operated until 1934.

Everett Hood, "umbrella mechanician and grinder of scissors and knives," travelled with his wagon throughout Vermont and New Hampshire. In this early-20th-century photograph, he has set up shop on North Park Street in front of the Carter fountain, with the Gerrish House (left, later torn down and replaced with a service station) and the Churchill House (right) on Campbell Street in the background.

Charles E. Pulsifer (1846–1922) and store clerk Bert Hapgood (1866–1919) pose in front of C.E. Pulsifer, Grocer, around 1889. The store operated for decades, passing to Charles's son Ernest after Charles's death. Pulsifer's was a quintessential Lebanon institution, and the building, located just west of the Whipple Block, was known as the Pulsifer Block long after the business closed in the 1920s.

This early 1930s photograph shows the interior of the First National Store, which moved into the Pulsifer Block after the original grocery store closed in the 1920s. This location of the First National Store was short-lived, but a sister store at 87 Hanover Street opened shortly thereafter and remained in operation into the 1980s.

In this 1914 photograph, Arthur J. Plamondon (1873–1964; right) shows off his new shipment of Maxwell automobiles in front of Colburn Park. Plamondon was born in Quebec, Canada, and moved to Lebanon as part of the 19th-century wave of Quebecois immigration to New England. In Lebanon, Plamondon was well-known for his many business ventures, including as an agent for the Maxwell Motor Company (later acquired by Chrysler).

Another of Arthur J. Plamondon's business ventures was the A.J. Plamondon shoe store. Plamondon ran his store out of the Billings Block at 9 Hanover Street, now located in the heart of Lebanon's pedestrian mall. In this c. 1915 photograph, Plamondon (left) poses with one of his employees in the midst of his wares.

Many of Lebanon's farms have remained in continuous operation since the 1760s, often within the same family. The Storrs-Townsend Farm, or Tomapo Farm, at the top of Storrs Hill is one such farm. Started in 1769 by Nathaniel Storrs (1747–1813), the farm was in the hands of his great-great-grandson Hugh Townsend (1888–1973) in 1915, when Townsend was photographed hauling timber. The farm remains in the family in the 21st century.

When the Rockland Military Academy closed in 1914, the school building was left vacant. West Lebanon took ownership of the building, which originally housed the Tilden Ladies' Seminary, and the school reopened for West Lebanon's elementary and high students the following year. This photograph shows one of the first classes in the new school, taught by Edith Bryant (1871–1963; standing at right).

Lebanon's central village did not have a dedicated postal service building before construction of the post office at 11 East Park Street in 1937. Until then, the post office resided in a variety of locations, first in Hough's Tavern (later known as the Lafayette Hotel), next in the town hall, and finally in the Hotel Rogers (shown here around 1922). Alonzo Chamberlin (1856-1926; center) was Lebanon's postmaster at the time.

Lander's Café and Sea Grill, shown here in 1935, opened at the corner of Mill and Hanover Streets in 1933. George and Michael Alafat ran the popular restaurant—the offshoot of John Landers's original institution—in the central village for three decades until it was destroyed in the 1964 fire. However, Lander's continued to be a Lebanon mainstay in a new location on Route 120 for another two decades.

Honey Gardens had its start in 1908, when Philip Townsend (1890–1975) began his beekeeping business, which he expanded into a full dairy in 1916. Located at 227 Mechanic Street, the original storefront (right) was built in 1924 and had to be relocated three times as Mechanic Street expanded. Honey Gardens was Lebanon's largest supplier of milk until the business was sold to Billings Farm in the 1970s.

In 1955, Honey Gardens delivered milk to more than 700 homes. Originally, these deliveries were performed by horse and wagon, but owner Philip Townsend converted to a fleet of trucks as soon as it was feasible. Arthur Clark (1920–2000) took over the dairy's operations from Townsend, his father-in-law, and is shown here grinning in the driver's seat. The back of this photograph bears Clark's proud commentary: "My first milk tanker, 1956."

In this 1961 photograph, two anchors of Lebanon's mid-20th-century commercial district stand side by side on Hanover Street: Woolworth's (at 15 Hanover Street) and McNeill's Drug Store (in the Perley Block at 9 Hanover Street). The Perley Block was torn down in 1963, after McNeill's relocated across the street. McNeill's remained a Lebanon mainstay until 1987, and Woolworth's closed in 1997 after eight decades in Lebanon.

The Blodgett Block stood at the corner of Court and North Park Streets for nearly a century before being demolished in 1950. Construction (shown here) immediately began on its replacement, the Mascoma Savings Bank building. The Harrison Block across Court Street (far left) was itself torn down in 1967 as part of urban renewal.

Bridgman's Furniture was started by Nathan Crossman Bridgman (1842–1922; seated with pipe) in 1891. It first opened in the basement of the Thompson Block on North Park Street (above) next door to Lebanon's town hall. As the business thrived and passed from one generation to the next, the storefront moved around Lebanon to larger and finer accommodations—first to the Bank Block on West Park Street in the early 20th century, and to a custom-built and expansive building on the Miracle Mile (below) in 1952. The business served Lebanon for over 125 years under the leadership of the Bridgman family until it finally closed in 2017.

Split Ballbearing had its humble beginnings on Mascoma Street at the foot of Benton Hill in 1927. World War II produced an exponential increase in the need for ball bearings that could be replaced without dismantling machinery, and "Split Ball" grew into one of Lebanon's most important industries. The combined Split Ballbearing and Miniature Precision Ballbearing moved into this new 30,000-square-foot factory on the Miracle Mile in 1958.

The Mary Hitchcock Memorial Hospital opened in Hanover in 1893 and served the Upper Valley's healthcare needs for nearly a century. By the 1980s, however, the limitations of the location became too much, and the hospital was moved to Lebanon in 1991. This photograph shows the new Dartmouth Hitchcock Medical Center under construction in May 1989.

Six

Lebanon at Play

Recreation occupied as much of the city's focus as its commerce and industry. Lebanon's residents engaged in a wide variety of leisure activities, from parades to sports, and documented them in copious photographs and written accounts. The Opera House hosted lectures and staged musical and theatrical performances, while the fields at Eldridge Park played host to Lebanon's athletes, including the Lebanon Senators baseball team. Colburn Park has long been the preferred site for outdoor gatherings and civic activities, while Lebanon's many schools offered venues for student performances and artistic endeavors in addition to sporting events.

Many of Lebanon's businesses supported the recreation and leisure of the city's residents in venues ranging from restaurants to resorts. Even the city's streets played host to community events, celebrating city anniversaries as well as events of national and global import. Across the broad category of recreation, the many locales highlighted in this chapter provided the backdrop for some of Lebanon's most enduring moments and memories.

Colburn Park has long been a center of leisure and entertainment in Lebanon. Here, a group of Lebanonians play croquet in the northern part of Colburn Park around the mid-19th century. North Park Street appears in the background, running in front of the Blodgett Block (left), the Thompson Block (center), and the original meetinghouse (right) before it was renovated in 1868.

The Chiron Spring House on Etna Road was "a summer hotel of rare excellence and attractiveness," according to an article printed in the *Boston Evening Transcript* in 1887. In addition to the "healthful" water produced by the nearby spring, the hotel offered 28 rooms, fine dining, a bowling alley, and reinvigorating outdoor activities. The hotel had a short existence, however, closing before 1900. By the 21st century, this had become an office building.

Members of Lebanon's banjo club—Clara Churchill at far left, and, in no particular order, Frances Hildreth, Mrs. DuBois, and Georgia Houghton Shaw—posed for this photo in 1893. In addition to playing, Clara Churchill also composed banjo music. In the same year, Churchill composed *Nonabel Schottische*, which was noted in publications both locally and nationally.

This domestic scene took place in the living room of Charles E. Lewis's (1844–1921) home at 32 Parkhurst Street in the late 1890s. Lewis was Lebanon's preeminent photographer, responsible for portrait photographs of many of Lebanon's residents during his career. Lewis applied his skills to his own family for this portrait of his second wife (seated) and their four children.

Nathan Crossman Bridgman (standing at center right), a longtime Lebanon auctioneer, is shown here enthusiastically overseeing an 1894 auction of furniture and household wares in Colburn Park. Three years earlier, Bridgman had expanded his business enterprise and opened a furniture store with his son Daniel (seated), who acted as clerk at this auction.

The Lebanon Choral Union performed the opera *Priscilla; or, The Pilgrim's Proxy* at the Lebanon Opera House in December 1895. The cast (pictured) included Fred G. Carter (far left), son of H.W. Carter, and Annie Currier (front right), a local music teacher. Tickets cost 35¢, and a special train ran from White River Junction to transport audience members to the performance.

The Lebanon Snowshoe Club enjoyed adventuring in the outdoors not only in Lebanon but throughout New Hampshire and Vermont. This February 1906 photograph commemorates the club's outing to Royalton, Vermont, where they stayed at the Cascadnac House (shown here). The Randolph, Vermont, *Herald and News* noted that the club also visited the newly constructed Joseph Smith Birthplace Memorial obelisk as part of their trip.

A popular form of election betting in the late 19th and early 20th centuries was the wheelbarrow bet, where the winner of the bet received a wheelbarrow ride from the loser. The unidentified men in this c. 1912 photograph paid off their election bet by transporting their passenger from the Mascoma Mill on Mechanic Street to White River Junction, Vermont, and back.

Eldridge Park, the athletic and recreational field located on the north side of Spencer Street, has long been one of Lebanon's predominant outdoor facilities. Originally known as Depot Meadow for its proximity to the Lebanon village train station, the site was deeded to Lebanon upon the death of its owner, Watson Eldridge. The land was largely given over to athletic fields and was named Rogers and Whitney Park in honor of two major supporters of the town's semiprofessional baseball team, the Lebanon Senators. Lebanon's high school teams also played here; the 1904 team is pictured below. The field was renamed Eldridge Park in the 1930s and continues to host sporting events well into the 21st century.

Lebanon celebrated its sesquicentennial in 1911, honoring the 150th anniversary of the 1761 town charter. This photograph shows Colburn Park, with West Park and Hanover Streets in the background, decorated for the occasion. A temporary speaker's stand was erected and surrounded by benches borrowed from the nearby town hall, and flags, bunting, and paper lanterns adorned the park and surrounding buildings.

Lebanon's residents and businesses competed to produce the most elaborate and eye-catching floats for the sesquicentennial parade held on July 4, 1911. Several dozen employees of the H.W. Carter and Sons factory participated in the float shown here awaiting the start of the parade next to the Densmore Brick Company barn at 166 Hanover Street.

Pres. William Howard Taft (1857–1930) was on vacation in New England when he passed through Lebanon. This photograph shows President Taft delivering a speech from his car to a crowd on North Park Street in October 1912. East Park Street is visible in the background, with the new Lebanon Public Library, built in 1908, visible at top left.

After the conclusion of World War I, Lebanon hosted a "Welcome Home Celebration" for returning soldiers in 1919. The centerpiece of the celebration was a parade; the James. B. Perry post of the Woman's Relief Corps is shown proceeding down North Park Street at the intersection with Court Street.

Ski jumping came to Lebanon via Erling Heistad (1897–1967), a Norwegian immigrant who moved to Lebanon in 1923. That same year, he founded the Lebanon Outing Club, and he soon built Lebanon's first ski jump at Gerrish Court. Four more followed, including jumps at Gerrish Farm (near Split Ballbearing), School Street, and Storrs Hill. This early photograph shows the sport's widespread popularity in Lebanon.

Starting in the 1930s, Doc's Place was a Mascoma Lake institution for decades. The snack bar and filling station, located on Route 4A near the Enfield town line, was accompanied by Doc's Cabins across the street, which hosted both locals and the lake's summer visitors. Doc's Place became the Baited Hook restaurant in the 1980s, and it continues to operate in the 21st century.

The Lebanon High School band, directed by music teacher Victor Wrenn (lower right), posed for this photograph around 1935 on the front steps of the Lebanon Public Library, located down the street from the new Lebanon High School on Bank Street (later the junior high school). Wrenn was appointed as the supervisor of music early in the 1934–1935 school year.

Victor Wrenn directed a 1935 production of the musical *Tulip Time*, a joint endeavor between Lebanon and West Lebanon High Schools. The production was staged at West Lebanon High School on Seminary Hill in the school's newly constructed state-of-the-art auditorium. Proceeds supported the West Lebanon High School Orchestra, shown here in the front row.

The 1943–1944 sports season represented the last opportunity for many Lebanon teenagers to engage in collegial athletic competition before joining the fighting forces in World War II upon graduation. Here, the West Lebanon High School boys' basketball team poses for their team photograph. Seven of the players would fight in World War II; happily, all returned home from their service.

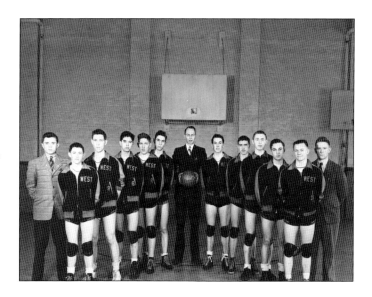

The Alice Peck Day Memorial Hospital opened in 1932 thanks to a bequest from Alice Peck Day (1860–1927). It began as a cottage hospital in the Peck family home, but the hospital so successfully served the Lebanon community that it soon required an expansion, which was completed in 1962. The community supported the hospital with its patronage and with Donation Days, like this one from the 1950s organized by the Lebanon Woman's Club.

Local newspaper the *Valley News* hosted a soapbox derby on August 6, 1954, as part of the Carter Community Building Street Fair. Twenty-seven children competed in three divisions, racing down Benton Hill on Mascoma Street in homemade vehicles capable of steering and braking. Though Mascoma Street still connects to Colburn Park, urban renewal in the 1960s dramatically altered this street and Benton Hill.

Lebanon recognized the 200th anniversary of its founding charter in 1961 with parades, dances, concerts, and other outdoor activities. The bicentennial celebration kicked off in West Lebanon with a parade, including the children's bicycle parade shown here lining up in front of the post office at 55 Main Street.

A second parade in the central village continued the bicentennial celebration on July 4, 1961. Floats competed for top honors and processed around Colburn Park and past the reviewing stand, shown here on the front steps of city hall. The judging panel was made up of members of the Bicentennial Committee and the Lebanon City Council.

Lebanon's first A&W Restaurant, pictured in the mid-1960s, opened on the Miracle Mile in 1962. This location operated for only seven years before moving to West Lebanon, but it defined a generation. The *Valley News* declared it "mecca to a generation of Upper Valley young people" upon its closure in November 1969. The restaurant offered both a sit-down dining room and seasonal carhop service.

Though the pedestrian mall remains one of the most controversial parts of Lebanon's history, the official opening of the mall on August 4, 1970, was a major community event. Hundreds of residents attended the ceremony, which included speeches by the governor and mayor Shirley K. Merrill (1922–1993), Lebanon's first female mayor, who is shown here preparing to speak.

The Four Aces Diner opened on the northeast corner of Main and Dana Streets in 1952. Manufactured by the Worcester Lunch Car Company, the diner (Lunch Car No. 837) served West Lebanon and Route 4 travelers until 1986, when this photograph was taken. One month later, the diner was moved one block west to 23 Bridge Street, where it continues to operate into the 21st century.

Seven

LANDMARKS

Every era of Lebanon's history produced notable buildings and monuments and was, in some way, defined by the notable landmarks of the time. Some landmarks were not manmade but rather natural features discovered and admired by Lebanon's residents. Some were built structures, highly prized and beloved by their creators and visitors, while others were primarily functional structures that would only be appreciated for their aesthetic or cultural significance in later years.

This chapter features landmarks from all of these categories, ranging from prehistory to Lebanon's colonial origins and through the 19th and 20th centuries. A landmark could be any structure or site of note—a peculiar geologic form, a house of worship, a memorial monument, an ornate residence, or a functional public facility. Many of these landmarks remain familiar sights around Lebanon, while others have been lost to time or memory. All of them, however, have held special significance for Lebanon and deserve a second look.

The First Congregational Church of Lebanon, shown here in 1911, was constructed in 1828 as Lebanon's first dedicated religious building. The elegant building was designed by local architect Ammi Burnham Young and sits on the corner of Church and South Park Streets. The church has been expanded over the years, but the exterior has remained largely unchanged since its construction.

The West Lebanon Congregational Church was built in 1849 and immediately became one of the centers of West Lebanon civic life. The parishioners had formerly been part of the Lebanon Congregational Church, but by the 1840s, the number of West Lebanon members justified the additional congregation. The church, shown here in an 1860s photograph by William W. Culver, was designed by Moses Wood (1822–1870).

Clara Churchill (left) and Ida Leavitt pose with the famous Tipping Rock in 1896. The Tipping Rock was a popular Victorian tourist attraction and picnic destination. The boulder balanced on one edge and would safely tip forward several inches with a little pressure. Visitors filled the space under the rock with potentially crushable items, and the rock—a victim of its own popularity—no longer tips.

The Devil's Kitchen, a collection of geologic potholes, was another popular natural attraction. Located in East Lebanon, these deep cylindrical depressions were created by thousands of years of water draining from the Glacial Lake Mascoma, the surface of which reached as high as 300 feet above that of the modern lake. The largest of the potholes could hold an entire person, as demonstrated by the man in this photograph.

After nearly a decade of fundraising and construction, the Soldiers Memorial on North Park Street was dedicated in 1890. The monument was built to honor veterans of the Civil War and originally housed Lebanon's public library on the first floor, with meeting space for veterans and their organizations on the second floor.

The Soldiers Memorial was elaborately decorated, with stained-glass windows honoring Lebanon's veterans and an assortment of sculptures flanking the entrance. The infantryman statue shown here was a gift of Frank Churchill, and the *Granite State Free Press* declared it "a splendid ornament" upon its dedication on Memorial Day in 1891. The other sculptures, added later, included cannonballs and artillery shells from the USS *Maine* and a cannon that stood guard over the entrance.

Colburn Park was named for Robert Colburn, who provided Lebanon with the land where the town's meetinghouse was built in 1792. The park was originally segmented by the intersection of the Fourth New Hampshire and Croydon Turnpikes. Lebanon rerouted the roads and fenced in the common around 1850, creating the dedicated park shown in this 1870 photograph looking toward School Street from North Park Street.

Known as the Rogers House since 1970, the elegant North Park Street building started as the Hotel Rogers, built in 1911. Architect John A. Fox (1835-1920) designed the hotel, which cost nearly $100,000 and had 44 rooms for 100 guests, as well as a restaurant. By the time of this 1940s photograph, however, the hotel was struggling to bring in guests, and the hotel closed in the late 1960s.

West Lebanon's library was started in 1869, but it took 40 years before it had a permanent home. The library moved between the West Lebanon schoolhouse, a local drugstore, and private residences until, after 25 years of fundraising, the West Lebanon Library (pictured) opened at 57 Main Street in 1909. The brick building served West Lebanon's residents until the Kilton Public Library opened in 2010.

Lebanon's library first resided in the Thompson Block at 55 North Park Street before moving to the first floor of the Soldiers Memorial. The library's first dedicated structure, however, was constructed in 1909—one of over 2,500 libraries funded by Andrew Carnegie. The land on East Park Street was donated by local businessman George Rogers. The library is shown here decorated for Lebanon's sesquicentennial celebration in 1911.

Henry R. Campbell (1807–1879) was the chief engineer of the Northern Railroad when he built his Lebanon residence on the corner of Bank Street and what was then Depot Street (later renamed in his honor). The house soon passed to Henry Wood Carter (1822–1897), founder of H.W. Carter & Sons, who would build his clothing factory next door (shown in the background).

Beside the Carter House stands the Churchill House, designed by John A. Fox for Frank and Clara Churchill. The Churchills travelled through much of the United States while Frank served as an agent of the US Department of the Interior. Clara hoped to turn the house into a museum for their collection of Native American art and artifacts, but the collection ultimately went to the Hood Museum of Art at Dartmouth College.

Lebanon's first Catholic church, established in 1856, was a converted storage building on Parkhurst Street. The first parish church, Lebanon's original Sacred Heart church, opened at 56 School Street in 1879. The church, shown above decorated for the 1909 visit of the church's former curate, Bishop George Albert Guertin, served Lebanon until it was replaced by its successor on High Street (pictured below). The new Sacred Heart Church was dedicated on July 14, 1942, and constructed of bricks from the Densmore Brick Company. The new location was closer to the Sacred Heart School on Eldridge Street and the convent next door, and was more convenient for parishioners, who were predominantly French-Canadians living in the neighborhood west of High and Hanover Streets.

The West Lebanon fire station was built at 63 South Main Street in 1893 and served West Lebanon for 80 years. By the 1970s, however, the station was in poor condition, with sagging floors and outdated facilities, and it was demolished in 1973. One of the fire engines in this 1964 photograph was purchased in 1963 for $15,000.

The Peck Homestead at 127 Mascoma Street was built by Simeon Peck (1733–1814) around 1780. The house was the Peck family home for generations until Peck's great-great granddaughter Alice Peck Day (1860–1927) had it converted via her will into a cottage hospital after her death. The Alice Peck Day Memorial Hospital remained in the Peck Homestead until 1964, when it expanded into a modern facility. The building remains part of the hospital campus.

Lebanon's city hall (originally the town hall) was built in 1923 after a devastating fire destroyed the previous town hall, which had served the community for 130 years. In addition to the town offices, the new hall included a theater, a banquet hall, and an auditorium that also hosted basketball games. The new hall was dedicated on October 29, 1924.

The Gulf station on the corner of North Park and Campbell Streets was built in 1952 and is shown here from the Carter House lawn in 1978. Its first 23 years were spent as the LaCoss Service Station, run by Raymond LaCoss (1913–1979), before passing to Roy Dickerson (born in 1932), who ran it as Roy's Auto Service until 2017. The building was then converted to a coffee shop, Lucky's Coffee Garage.

Eight

LEBANON ON FIRE

In the era of predominantly wooden structures and before the development of modern fire suppression and firefighting infrastructure, fires were strikingly common. Lebanon seems to have had more than its fair share of fires between 1887, the year of Lebanon's first "great fire," and 1964, the year of its second "great fire."

Destructive conflagrations could be caused by unattended lanterns, poorly maintained heating infrastructure, faulty electrical wiring, inattentive smokers, or, as in two notable Lebanon cases, arson. The density of structures in Lebanon's villages meant that a fire rarely stuck to its place of origin; if one building burned, it often took a second structure—or many more—with it. The intermingling of industrial, commercial, and residential buildings within Lebanon's dense villages also contributed to the frequency and intensity of its fires. A residential fire would often burn out quickly, but a fire that reached wood shavings in a sawmill or industrial chemicals in a factory was much more difficult to contain.

Lebanon's era of fires largely came to an end with the rise of brick and steel construction and the development of modern fire suppression and firefighting infrastructure, and the city has witnessed few truly destructive fires since the 1964 conflagration. The legacy of Lebanon's fires, however, is still evident in the surviving brick industrial buildings, the downtown transformed by urban renewal, and the varied architectural styles throughout the city that attest to Lebanon's ongoing building and rebuilding.

By the final decades of the 19th century, Lebanon was a thriving industrial and commercial center. This picture, taken from Lebanon's town hall tower around 1875, shows the commercial buildings and mills on Benton Hill leading down to Mechanic Street and Mascoma Street at upper right. The Carter and Rogers factory (later the Lebanon Woolen Mill) and its distinctive cupola are visible at upper left.

Though fires were common in 19th-century Lebanon, the one that raged on the corner of Mascoma and West Park Streets on February 10, 1879, was deemed Lebanon's Great Fire—until it was overshadowed by a much greater fire only eight years later. The 1879 fire destroyed the Durant & Perkins furniture company (shown at far left in the previous photograph), the Odd Fellows hall in the Worthen Block, and four residences.

Just after midnight on May 10, 1887, an overturned lantern in the Mead, Mason and Company shop on Water Street ignited the factory. The conflagration swept through town, destroying every structure in its way. The gutted structure at far left was the foundry at 1 Foundry Street, which is still standing in the 21st century. The fire was finally halted at High Street, which is visible in the background.

Forty families from single-family homes and apartment buildings were displaced by the 1887 fire. Many of their hastily evacuated possessions, loaded into wagons and sleighs, found safety in Colburn Park alongside the contents of the liveries, gristmills, machine shops, grocers, cobbler shops, and photography studios destroyed by the disaster. Remarkably, no lives were lost in the fire.

This view from Mechanic Street shows the destruction of nearly every building from the first photograph in this chapter. By the time the 1887 fire was extinguished, 12 acres and nearly 80 structures had been reduced to rubble, including 20 mills and manufacturers. The damaged area stretched east from High Street to West Park Street, stopping just short of the newly constructed Whipple Block and Colburn Park, and south from Hanover Street to the southern end of Foundry

Street. The surviving buildings shown here include the Whipple Block (left), the town hall, the new Odd Fellows Hall (rebuilt after the 1879 fire), and the Lebanon Congregational Church (right, with a white steeple). In the foreground are the ruins of the Lebanon Woolen Mill (bottom right), the Kendrick & Davis factory, the Baxter machine shop, and Mead, Mason and Company—where the fire originated.

The Whipple Block was constructed by Gilman C. Whipple (1837–1910) in 1882 on the corner of Hanover and West Park Streets. It was one of the few brick structures in Lebanon, constructed of local brick and stone from the Mount Support quarry. By 1894, the building was home to Richardson and Emerson's grocery, the I.N. Perley drugstore, and the Masonic hall on the third floor.

While the Whipple Block escaped the 1887 fire, it was less fortunate on January 15, 1894. A fire began in the basement and gutted most of the structure. The damage appeared worse than it was, however, and the building was quickly repaired; owner Gilman C. Whipple installed electric lights the following month, and all of the business occupants had reopened by that summer.

The Hildreth (left) and Simmons Blocks stood across Hanover Street from the Whipple Block. Hildreth's Hardware was established in 1855 and moved into this building on the north side of Hanover Street around 1880. The Simmons Block was home to a number of grocery stores, including Moulton & Freeman in the late 19th century and W.E. Mudgett's grocery in the early 20th century.

Early on January 10, 1903, a fire broke out in W.E. Mudgett's grocery in the Simmons Block. It immediately spread next door to the Hildreth Block. Firefighters decided that the two buildings were a lost cause and focused on saving the neighboring structures. Their efforts fighting the fire in subzero temperatures were successful, and only the Simmons Block and Hildreth Block (shown here from Hanover Street) were lost.

Lyman Whipple (1834–1908) (no relation to Gilman) built his construction shop on the corner of Mill and Mascoma Streets in 1888 after the previous shop was destroyed in the 1887 fire. Whipple was responsible for the construction of many of Lebanon's structures in the following decade, including Everett Knitting Works and the stables at the Riverdale Park track.

By 1904, Lyman Whipple had retired, and his shop was rented to Freeman & Grow, box-makers. On March 31, 1904, the shop caught fire, causing approximately $6,200 of damage. Whipple rebuilt the shop, renting it to the Marston Rake Company after their factory on Water Street burned in 1906. The reconstructed shop stood into the 1960s, when it was torn down as part of urban renewal.

Lebanon's first roller-skating rink—shown here from Hanover Street, in the left foreground—opened in May 1884. The building was constructed on Taylor Street next to Eldridge Park. The rink surface was made of birch, with viewing galleries capable of seating 250. Attached to the exterior of the rink were grandstands for the neighboring Rogers and Whitney Park.

The end of the roller-skating rink came less than 25 years later, on December 14, 1907. The structure caught fire early on that Saturday morning, and strong winds contributed to the loss of the entire rink. The attached grandstands were also severely damaged. The grandstands were reconstructed, but the skating rink was never rebuilt.

By 1923, Lebanon's town hall—formerly the meetinghouse originally constructed in 1792—had been the center of Lebanon's civic activity for over 130 years. After extensive renovations in 1868 and 1899, the Victorian-style town hall was the centerpiece of North Park Street, nestled between the Thompson Block and the new Hotel Rogers.

The town hall was entirely destroyed by fire on February 2, 1923. The cause of the fire is unknown, but the building, which—despite its many renovations—was still constructed of 18th-century wood, burned swiftly. The fire also devastated the Thompson and Kendrick Blocks to the west (shown here) and damaged the west end of the Hotel Rogers.

This picture taken from the rear of the former town hall shows the ruins with Colburn Park in the background. Fortunately, the town's vital records were kept in a brick vault that escaped the blaze unscathed and is shown here at center. Also visible are the brick foundations and columns used to lift the town hall during its 1899 renovation.

The original Lebanon fire station stood on the south side of Hanover Street, just east of the dry bridge over the railroad. This photograph from around 1930 shows Lebanon's firemen with their three fire engines. While it always offered an important public service, Lebanon's fire department would prove particularly important over the following two decades.

The Lebanon Excelsior Mill was built in 1903 in Riverdale on the south side of the Fourth New Hampshire Turnpike (now Bank Street Extension). Using the Mascoma River for power, it produced excelsior—wood shavings used for packing material and furniture stuffing. A railroad siding (pictured) connected the mill to the Northern Railroad roughly one mile east of the Lebanon central village.

On July 22, 1924, the Lebanon Excelsior Mill caught fire and burned to the ground. The *Granite State Free Press* reported that the town's new motorized water pump was able to save several thousand cords of wood stacked beside the mill, but the building was a complete loss. This photograph was taken looking southwest from what had been the interior of the mill with the nearby railroad bridge in the background.

The Park Hotel and the Odd Fellows Hall stood on West Park Street at the top of Benton Hill, roughly where Mascoma Street now runs, and across the street from the Bank Block. The Park Hotel was built in 1902, while the Odd Fellows Hall was constructed around 1885 and contained several businesses, including the offices of the *Granite State Free Press*.

Both the Park Hotel and the Odd Fellows Hall were destroyed by fire on February 11, 1931, in the midst of the Great Depression. The cause of the fire was unknown until the remains of the hotel's owner, James Venetsanos, were found in the basement—beside a can of gasoline. His car had been left idling outside, but he had failed to escape the fire and collect the insurance payout.

After surviving the 1894 fire, the Whipple Block had become the anchor of the Hanover Street commercial district. By the 1930s, the building was home to nearly one dozen businesses, including the Richardson and Langlois clothing store, Freeman and Davis jewelers, insurance and lawyer's offices, and milliners and electronics shops.

Guests in the Hotel Rogers on North Park Street first spotted the smoke rising from the Whipple Block on New Year's Day in 1932. The fire started in the boiler room and spread through the building, predominantly between the walls and floors. The damage displaced the building's many occupants, including—once again—the Masonic hall. However, as in 1894, the damage was repaired, and the building was reopened by that summer.

The Harrison Block was constructed at the northwest corner of Colburn Park by G.W. Worthen in 1871. The building later passed to J.E. Lincoln and his dry goods shops, and ultimately was inhabited by Frank B. Harrison and his clothing shop. By the time of this 1932 photograph, the Harrison Block was one of the oldest buildings on Hanover Street.

On November 21, 1933, a fire tore through the Harrison Block, causing an estimated $50,000 in damage. Victims included clothing stores on the ground floor, legal offices and a barbershop on the second floor, and apartments on the third floor. The damage was limited, however, and the building was restored as a two-story building. The Harrison Block stood until 1967, when it was demolished during urban renewal.

West Lebanon High School on Seminary Hill began its life as the Tilden Ladies' Seminary in 1855. The Rockland Military Academy (pictured) took over the building in 1903 and operated until 1914. The following year, the school reopened as West Lebanon High School, replacing the overburdened school on Main Street.

On February 19, 1940, a fire broke out in the oldest part of West Lebanon High School. Firefighters from West Lebanon, Lebanon, and White River Junction fought the inferno. Their efforts contained the fire to the old wing of the school and, fortunately, left the newly constructed auditorium and gymnasium largely undamaged.

The main part of West Lebanon High School, which was originally constructed in 1855, was entirely gutted by the 1940 fire, leaving only the exterior brick walls intact. The high school's records, books, and other educational materials were all destroyed. Students were sent to Lebanon High School for the remainder of the 1939–1940 school year while repairs were made.

The exterior walls of the damaged part of the school were deemed structurally sound and incorporated into the renovated school, which reopened in 1940. The building continued to house West Lebanon's high schoolers until 1961, when Lebanon and West Lebanon's residents voted to close the school and consolidate the two high schools.

A fire broke out in the basement of the Woolworth Block before dawn on January 12, 1945. Nestled in the middle of the Hanover Street commercial district, the fire threatened to spread into the neighboring buildings housing McNeill's Drug Store and Conti's Restaurant. After fighting the conflagration for over four hours in subzero temperatures, firefighters extinguished the fire, which caused no more than smoke damage to the neighboring buildings.

The damage to the Woolworth Block itself was estimated at $90,000, and the building was declared a complete loss. Construction of a new Woolworth's building began soon after the fire. Though the fire did not claim any lives, two men were killed two months later when a wall collapsed during the demolition of the remaining shell of the building.

By the middle of the 20th century, Lebanon was a bustling town—and, after a 1957 referendum that changed Lebanon's form of governance, an equally bustling city. Hanover Street (at center left in this mid-century photograph) was Lebanon's commercial hub, serving customers from Lebanon and neighboring towns. Though the city's woolen industry was in decline, new manufacturing was on the rise, and Lebanon's population consistently increased throughout the middle of the century.

The end of the Hanover Street commercial hub came on the afternoon of June 19, 1964, when two young men set fire to a rug in an abandoned building on Mill Street. The fire would ultimately destroy 20 buildings, displace 20 businesses and 100 residents, and take the lives of two people. The aftermath of Lebanon's second great fire is shown in this aerial view of the area.

The 1964 fire began in an abandoned blacksmith shop on Mill Street and quickly spread down the street and into the buildings on Hanover Street. The burning rubble on the left side of this photograph was all that remained of the east side of Mill Street; at right, the Currier & Company fabric store and Endicott Johnson Shoe Store are engulfed in flames.

More than 150 firefighters from Lebanon, neighboring towns, and communities as far away as Sunapee, New Hampshire, and Woodstock, Vermont, came to fight the 1964 fire. Here, fire hoses are directed at McNeill's Drug Store and Western Auto, housed in the Pulsifer Block. McNeill's was entirely lost, but Western Auto survived thanks to the firewall installed in the Pulsifer Block.

Word of the 1964 fire spread quickly, and residents of Lebanon and nearby towns came to watch the inferno as firefighters and volunteers worked to contain it. This photograph shows the intersection of Court, North Park, and West Park Streets, all of which were closed to traffic to facilitate the firefighting efforts but full of spectators spilling over from Colburn Park (at lower right).

By the time the 1964 fire was extinguished, 20 buildings were destroyed, and Lebanon had been declared a federal disaster area. This photograph, taken from the roof of the Whipple Block, shows the destruction northwest along Hanover Street, including the collapsed dry bridge over the railroad and the ruins of Lander's Restaurant, Lewis Brothers' Hardware, and many other Hanover Street storefronts.

The Lebanon United Methodist Church was built on School Street in 1833 and was a remarkable architectural achievement for the town's Methodist community after decades spent meeting in private residences, schoolhouses, and even barns. The church was enlarged and updated several times over its 160-year existence, with the final renovation occurring in 1969.

The Methodist church was accidentally set on fire by trespassers on February 21, 1992, and the fire quickly consumed the historic structure. Despite the loss of a century and a half of history and décor, including 19th-century stained-glass windows, the community banded together to rebuild the structure. The new United Methodist Church was consecrated in March 1994.

BIBLIOGRAPHY

Brooks, Lisa, Donna Roberts Moody, and John Moody. "Native Space." In *Where the Great River Rises: An Atlas of the Connecticut River Watershed in Vermont and New Hampshire*, ed. Rebecca A. Brown, 133-137. Lebanon, NH: Dartmouth College Press, 2009.

Carroll, Roger. *Lebanon, 1761-1994: The Evolution of a Resilient New Hampshire City*. West Kennebunk, ME: Phoenix Publishing, 1994.

Daniell, Jere. "Early Settlement." In *Where the Great River Rises: An Atlas of the Connecticut River Watershed in Vermont and New Hampshire*, ed. Rebecca A. Brown, 138-144. Lebanon, NH: Dartmouth College Press, 2009.

Downs, Charles A. *History of Lebanon, N.H., 1761-1887*. Concord, NH: Rumford Printing Co., 1908.

Labbe, Matthew, and Robert Goodby. "The Water Powered Mills of Lebanon, New Hampshire." Stoddard, NH: Monadnock Archaeological Consulting, LLC, 2018.

Leavitt, Robert Hayes. *Lebanon, New Hampshire In Pictures, Vols. I-II*. Lebanon, NH: Whitman Publishing Co., 1997.

Mathewson, R. Duncan, III. "Western Abenaki of the Upper Connecticut River Basin: Preliminary Notes on Native American Pre-Contact Culture in Northern New England." *The Journal of Vermont Archaeology* 12 (2011): 1-45.

Millen, Ethel Rock. *Historical Sketches of Early Lebanon, New Hampshire*. Canaan, NH: Reporter Press, 1965.

Papazian, Lyssa. "Colburn Park Historic District Update—Phase I Report." Putney, VT: Self-published, 2021.

Torbert, Edward N. "The Evolution of Land Utilization in Lebanon, New Hampshire." *Geographical Review* 25:2 (Apr. 1935): 209-230.

Discover Thousands of Local History Books Featuring Millions of Vintage Images

Arcadia Publishing, the leading local history publisher in the United States, is committed to making history accessible and meaningful through publishing books that celebrate and preserve the heritage of America's people and places.

Find more books like this at
www.arcadiapublishing.com

Search for your hometown history, your old stomping grounds, and even your favorite sports team.

Consistent with our mission to preserve history on a local level, this book was printed in South Carolina on American-made paper and manufactured entirely in the United States. Products carrying the accredited Forest Stewardship Council (FSC) label are printed on 100 percent FSC-certified paper.

MADE IN THE